环保卫士

小猪爱可

谁把暖气打开了

小猪爱可解释全球变暖

[美] 丽莎·弗兰奇 著　[美] 巴瑞·戈特 绘　张玉亮 译

江西科学技术出版社

Original title: ECO-Pig Explains Global Warming: Who Turned Up the Heat

图书在版编目（CIP）数据

谁把暖气打开了：小猪爱可解释全球变暖：英汉对照 / (美) 弗兰奇著；(美)戈特绘；张玉亮译.
-- 南昌 :江西科学技术出版社, 2016.8
（小猪爱可讲环保）
ISBN 978-7-5390-5488-9
Ⅰ. ①谁… Ⅱ. ①弗… ②戈… ③张… Ⅲ. ①环境保护 - 少儿读物 - 英、汉 Ⅳ. ①X-49

中国版本图书馆CIP数据核字(2016)第026837号

国际互联网（Internet）地址：http://www.jxkjcbs.com
选题序号：KX2016083　　图书代码：D16001-101

小猪爱可讲环保
谁把暖气打开了：小猪爱可解释全球变暖

文 / (美)丽莎·弗兰奇　图 / (美)巴瑞·戈特　译 / 张玉亮
责任编辑 / 刘丽婷　美术编辑 / 刘小萍　曹弟姐
出版发行 / 江西科学技术出版社
社址 / 南昌市蓼洲街2号附1号　邮编 / 330009
电话 / (0791)86623491　86639342(传真)
印刷 / 江西华奥印务有限责任公司
经销 / 各地新华书店
成品尺寸 / 235mm × 205mm　1/16
字数 / 50千　　印张 / 8
版次 / 2016年8月第1版　2016年8月第1次印刷
书号 / ISBN 978-7-5390-5488-9
定价 / 50.00元（全4册）

On top of a mountain
at the edge of a sea
lives an Earth-loving pig
in a town called To-Be.

在海之滨、山之巅的未来小镇上，生活着一只热爱地球的小猪。

His name is Bernard
but please call him E.P.,
should you visit his home
in the green apple tree.

他叫佰纳德，小名叫爱可。他住在一棵青苹果树上。如果你去他家里参观，你就会知道为什么大家都叫他“环保卫士”了。

From his Eco-Pig nest,
tenth branch from the bottom,
he keeps an eye on our planet,
winter, spring, summer, and autumn.

从树干底部往上数到第十根树枝，你就能发现小猪爱可的家了。在那里，他时刻关注着我们的地球，从春天到夏天，从夏天到秋天，从秋天到冬天，再从冬天到春天……

On the first day of March,
not quite yet spring,
E.P. was recycling,
his most favorite thing.

3月的第一天，春天的脚步还没走近，小猪爱可正忙着废物回收利用。这可是他最喜欢做的事情啦！

As he sorted out cans
he stopped and he thought,
My goodness, my snout
is unusually hot.

在将垃圾罐进行分类时，他突然停了下来：我的天啊，我的鼻子怎么这么热啊？

It was hot to the touch
and blinking bright red.
So were the ears
on top of his head.

他的鼻子摸上去很烫，而且通红通红的。他的两只耳朵也变得像鼻子一样又热又红。

A sunburn in March?
How could this be?
It just snowed yesterday.
Earth is singing off-key.

才3月就会被晒伤吗？怎么会这样呢？最近才刚刚下过雪。地球太反常了。

"And have you noticed," cried Lou,
"the tulips are out,
and the red-breasted robins
are hopping about?

"你注意到了吗？"洛尔喊道，"郁金香已经开花，红胸知更鸟也出来活动了。"

"From winter to summer?
This much I know,
there should be four seasons.
Now where did they go?"

“从冬天直接到夏天了？据我所知，应该有四个季节啊。现在，那些跳过的季节都到哪儿去了？”

"Earth's running a fever," E.P. agreed.
"We've cranked up the heat!
It's called global warming,
so not cool and not neat.

"地球发烧了，"小猪爱可点头赞同道，"这就是人们常说的全球变暖，我们已经加快了地球变暖的速度！所以地球现在已经不凉爽，也不干净了。

"We turn on all our lights,
and we drive and we fly.
But I'm starting to wonder
what we've done to the sky.

"我们让所有的灯都亮着，我们还驾驶着各种车辆在路上跑，乘着各种飞机在天上飞。我们是时候该考虑考虑对天空造成的影响了。

"When we power the planet
we need to be fair!
Burning oil and coal
puts greenhouse gas in the air.

"在我们努力建设我们的地球家园时，也需要遵守大自然的法则。燃烧石油和煤炭会向空气中排放温室气体。

“Some gas gets swallowed
by our oceans and trees.
But we’re making too much
and they’re starting to wheeze.

“有些废气被海洋和树木吸收了。但是我们排放了太多的废气，海洋和树木也开始气喘吁吁、无力应对了。

"This bad gas we can't see
traps the heat from the sun.
Then Earth gets too hot,
like a grilled burger, no bun.

"这些我们无法看到的有害气体使来自太阳的热量无法消散。地球变得越来越热，就像在烘烤一个没有面包片的汉堡包。

"If Earth's temperature rises
by just one small degree,
that means really big changes
for the land and the sea!"

"虽然地球的温度只是升高小小的1℃，但这会对大地和海洋产生很大的影响。"

"Just one degree?" asked Lou.
"Is that such a big deal?
Are you sure, E.P.?
Is this warming for real?"

"只是升高1℃？"洛尔问道，"就会对地球产生很大影响吗？你确定吗，小猪爱可？这种变暖是真的吗？"

E.P. replied, "Keeping the climate
in balance is part of nature's big plan.
Everyone knows
pigs aren't meant to be tan!

小猪爱可回答说："保持生态平衡是动植物生存大计的一部分。大家都知道，我们这样粉嫩的小猪可不希望被晒黑！

“Plus, glaciers are melting,
which puts our friends underwater,
and that sure doesn’t work,
unless you’re an otter.

“此外，冰川正在融化，快把我们的北极熊朋友逼入水下了。这样肯定是不行的，除非它们能变成海獭。

"While some places flood," E.P. cried,
"others will fry.
Our fruits and our veggies
could just curl up and die.

“当有些地方暴发洪水时，”小猪爱可大声说，“另一些地方却可能遭受着炎炎烈日的炙烤。我们的瓜果蔬菜只能蔫头耷脑地枯萎。

"To get Earth back in tune,
we need to go Green.
We need to start now,
if you know what I mean!"

“为了让地球重回和谐，我们需要变得‘绿色环保’。如果你懂我的意思，那我们就需要从现在做起！”

Electric power and cars
warm the planet the most.
If we don't use them less,
then our climate is toast!

电力和汽车是让地球变暖的罪魁祸首。如果我们依然过度使用电和汽车，恐怕就得生活在像烤箱一样热烘烘的“圆形大熔炉”里了！

If you don’t have to go far,
ride the bus or your bike.
Need to go to the store?
Take your dog for a hike!

如果你去的地方不远，请乘坐公交车或骑自行车。如果你想去商场，请带上你的狗狗一起溜达过去吧！

For a major road trip,
why not share a ride?
You'll make some new friends,
and show your Earth pride.

如果你要旅行，你可以选择拼车。这样你不仅可以交到新朋友，还可以和他们一起分享保护地球带来的自豪感。

We can learn to make power
from the wind and the sun.
It's renewable and clean.
Plus, wind farms are fun!

我们可以学习利用风能和太阳能发电，这些是可再生的清洁能源。此外，风力发电厂可是非常有趣的哦！

Let's switch off and unplug
the things we don't need.
We'll reduce greenhouse gas.
On this we're agreed!

　　让我们关闭电源，拔下我们不需要的电器插头吧。这样做就可以减少温室气体排放。**我们都愿意这么做！**

If we each cut back a little,
it will add up to a lot.
Let's pitch in and help out
before our home gets too hot!

如果我们每个人为环保贡献一点儿，地球就会变得更美好一些。在我们的地球家园变得太热之前，让我们共同携手，帮助它走出困境吧！

Let's give back to our planet,
not always subtract.
Because what we do matters,
and that is a fact!

让我们回报地球，而不是一味地向它索取，破坏它的生态平衡。我们的行为与地球的生机息息相关。这就是事实！

必学词汇

全球变暖——地球温度逐渐增高，进一步引发气候变化。

绿色环保——与环境或保护环境相关的（事物）。

温室气体——将太阳的热量存留在大气中的气体。

回收利用——将废物、玻璃或易拉罐分类，以便再次利用。

可再生能源——由自然资源生成的能源，如太阳能和风能。

风力发电厂——将风能转化为电能的工厂。

你知道吗？

- 地球的温度在上个世纪上升了0.56℃。科学家预计地球的平均温度在这个世纪会再上升1.11~3.33℃。
- 气候变化会影响海平面、庄稼、空气和水。
- 剧烈的气候变化无法给地球上的动植物足够的时间去适应环境，使一些物种有濒临灭绝的风险。
- 由于全球变暖，热浪、强热带风暴和野火出现的次数也在增加。
- 对抗全球变暖的一种方式是减少化石燃料的使用。化石燃料，如煤炭和石油，是由史前动植物形成的。风能和太阳能是由自然资源转化而来的能源，它们对我们的环境更有利。

减缓全球变暖的更多方法

与你的父母聊聊你们在家中可以做些什么

1. 拼车上学和参加课外活动。
2. 使用公共交通工具。
3. 骑自行车或步行至较近的地方。
4. 回收纸张、塑料、玻璃和金属。
5. 清理自己的房间，并将较新的衣物和玩具捐赠给有需要的人。
6. 当你不看电视，不使用电脑时，请把它们的插头拔下来。
7. 离开房间时记得关灯。